Simara Saquet Schio

Environmental awareness

Simara Saquet Schio

Environmental awareness

An Inference with Rural and Urban Families Linked to the E. E. M. Érico Veríssimo School (Restinga Sêca, RS, Brazil) E. E. M. Érico Veríssimo School (Restinga Sêca, RS, Brazil)

ScienciaScripts

Imprint

Cover image: www.ingimage.com

This book is a translation from the original published under ISBN 978-3-330-76679-2.

Publisher:
Sciencia Scripts
is a trademark of
Dodo Books Indian Ocean Ltd. and OmniScriptum S.R.L publishing group

120 High Road, East Finchley, London, N2 9ED, United Kingdom
Str. Armeneasca 28/1, office 1, Chisinau MD-2012, Republic of Moldova, Europe
Managing Directors: Ieva Konstantinova, Victoria Ursu
info@omniscriptum.com

Printed at: see last page
ISBN: 978-620-8-54436-2

ACKNOWLEDGEMENTS

Firstly to God, for giving me enough gifts to *get me this far. I know that "I can do all things through him who strengthens me".*

I would like to thank my family for their dedication, encouragement and love.

To my son Patrick and my husband Josemar for all their love, support, patience and understanding when I often had to give up our time together to do this work. I am eternally grateful.

To my supervisor, Professor Jumaida Maria Rosito, for her dedication, patience and encouragement at every stage of this work.

The new ecological awareness and the notion of the Earth as a great mother and Gaia involves a pedagogical process through which people become sensitised to this reality. Concepts are not enough. We need emotions, because they are what mobilise action. (Leonardo Boff, 2000)

SUMMARY

The environmental issue has become one of the most debated topics of the moment. Population growth, the production model and the unequal consumption of the planet's inhabitants are almost in conflict with humanity's quality of life, the conservation of physical environments and the integrity of organisms. Environmental issues are becoming increasingly urgent and important for society. With this in mind, this study aims to infer the degree of environmental commitment of rural and urban families linked to the Érico Veríssimo State High School in Restinga Sêca (RS). In this study, we combined bibliographical research with field research in order to, firstly, gather theoretical information on the subject so that it can serve as a basis for the second stage, which will be the field research, which used a semi-structured questionnaire covering environmental issues, especially the disposal of household waste, to be answered by students in the 1st, 2nd and 3rd grades of high school at the school. The questionnaires were evaluated and compared according to the students' origins (rural or urban), both qualitatively and quantitatively. This research is therefore justified.
The aim is to encourage greater reflection on environmental issues and the importance of raising awareness and sensitisation regarding the disposal of household waste by all individuals, whether they live in rural or urban areas.

Keywords: Environmental issues. Household rubbish. Raising awareness.

SUMMARY

CHAPTER 1

INTRODUCTION

The environmental issue has become one of the most debated topics of the moment. Society is suffering an intense crisis, which cannot be described as environmental, but rather as civilisational (BRAGA, 2003). Population growth, the production model and the unequal consumption of the planet's inhabitants are almost in conflict with humanity's quality of life, the conservation of physical environments and the integrity of organisms. Environmental issues are becoming increasingly urgent and important for society, as the fate of humanity depends on the relationship established with nature and the use of its available natural resources (BARRETO, 2006).

Since man's ancestors first appeared on earth, nature has been transformed. For a long time, man ensured his survival by causing little interference with the fauna and flora, as he hunted and gathered only what he needed for his daily maintenance. The use of fire was one of man's first great discoveries. Perhaps this was the first environmental impact caused by mankind. Some civilisations began to grow food, raise animals, develop new techniques, improve their knowledge and, consequently, expand their mastery of nature.

> [...] as the human species has developed new technologies and expanded its dominion over the elements and nature in general, environmental impacts have increased in intensity and extent (BRANCO, 1997, p. 20).

With the agricultural revolution, the impact on nature gradually began to increase as forests were cut down to create crops and artificial pastures. Over time, around the 18th and 19th centuries, the industrial revolutions profoundly changed the lifestyle of a large part of the population. In addition to population growth, there were advances in production techniques and the circulation of goods, increasing the capacity of human beings to transform nature.

As a result, environmental impacts have grown at a rapid pace, causing imbalances not just locally, but on a global scale. Such is the aggression and intensity caused by human influence that nature is unable to regenerate itself.

From this perspective, it can be inferred that human beings must realise that they are an integral part of nature and not superior to it, that without the environment and its resources, our species cannot survive. As Branco (1997, p. 21) emphasises, man, whether he likes it or not, depends on the existence of a rich, complex and balanced nature around him". So it's not enough to keep species alive, in order to preserve the balance between human beings and nature, it's necessary to make conscious use of natural resources, prioritising development without affecting the quality of life of future generations.

Nowadays, the majority of the population lives in urban centres. Clean water comes out of the tap and dirty water goes down the drain. The rubbish produced every day is taken away from the front of houses without people having the slightest concern about where it goes. In other words, the vast majority of people fail to realise the close correlation between the environment and their daily lives (DONELA, 1997).

This highlights a problem that concerns the formation of values and responsibility in the context of a crisis that affects all societies around the world. It is a structural and organisational crisis of civilisation, built up over the last few centuries and which is now globalised. This crisis has reached accelerated and acute levels and today forces the population to think and find innovative solutions to deal responsibly with the demands and challenges that reality presents, which involve both the present and the future of humanity.

It is clear that it is important to sensitise human beings to act responsibly and with awareness, maintaining a healthy environment in the present and for the future, so that they know how to demand and respect their own rights and those of the entire community, both local and international.

In this sense, the school plays a fundamental role in building the environmental awareness of students, who take on the role of a central element in the teaching and learning process, actively participating in the diagnosis of environmental problems and in the search for solutions, through the development of skills and the formation of attitudes, leading to ethical behaviour, consistent with the exercise of citizenship. According to Sato (2004), environmental learning is a vital component, as it provides reasons for students to recognise themselves as an integral part of the environment in which they live and makes them think about alternatives for solving environmental problems, as well as helping to maintain resources for future generations.

However, the school should not act alone in this environmental awareness campaign; it needs to be allied to the family, which plays a decisive role in formal and informal education. It is at the heart of this space that ethical and humanitarian values are absorbed and where bonds of solidarity are deepened. It is also within this space that the marks between generations are built and cultural values are observed.

In the same vein, we understand that a child who learns at school the importance of sorting rubbish for recycling, for example, will possibly pass it on to other family members, demanding that they take a proactive stance in this regard. One of the most effective ways of promoting and encouraging environmental education can and should take place in the most intimate spheres of human relationships, i.e. family life. Authors such as Dessen and Polonia (2007) understand that "the family and the school emerge as two fundamental institutions for triggering people's evolutionary processes, acting as drivers or inhibitors of their physical, intellectual, emotional and social growth".

In view of the above, this study aims to infer the degree of environmental awareness of rural and urban families linked to the Érico Veríssimo State High School in Restinga Sêca (RS). To this end, semi-structured questionnaires covering environmental issues, especially the disposal of household rubbish, were administered to students in grades 1, 2 and 3 at the school. The questionnaires were evaluated and compared according to the students' origins (rural or urban), both qualitatively and quantitatively.

CHAPTER 2

LITERATURE REVIEW

2.1 A brief consideration of Environmental Education

Environmental problems such as floods, landslides, avalanches, droughts, rising temperatures, among others, experienced by the Brazilian population, or even worldwide, are increasingly frequent and have worrying consequences. According to Marcatto (2002), the population, which is often the cause and victim of some of these environmental problems, is in a position to diagnose the situation and is interested in resolving it.

According to this author, the active and democratic participation of the local population in all phases of the process, which starts with discussing the problem, diagnosing the local situation, identifying possible solutions, implementing the alternatives and evaluating the results, is fundamental for a paradigm shift to occur.

In this sense, environmental education has emerged as a tool to help raise awareness and train the population about environmental problems. According to Carvalho (2006), environmental education initially arose from the concern of ecological movements to raise awareness of the importance of the conscious use of natural resources and became better known in the 80s and 90s. Internationally, this issue began to be discussed in 1972, at the first United Nations World Conference on the Human Environment, which took place in Stockholm, Sweden. On this occasion, the Stockholm Declaration drew up a set of principles for the ecologically rational management of the environment.

Environmental education is addressed by Brazilian legislation, more specifically by the National Environmental Education Policy, Law No. 9.795, of 27 April 1999. According to this legislation, education is seen as:

> [...] the processes through which individuals and the community build social values, knowledge, skills, attitudes and competences aimed at conserving the environment, which is an asset for the common use of the people, essential to a healthy quality of life and its sustainability (BRASIL, 1999).

In a complementary way, Medina (2000) characterises environmental education as a process that provides a critical and global understanding of the environment, in which it is necessary to transmit values and develop attitudes, which make it possible to adopt conscious and participatory positions with regard to the conservation and use of natural resources, improving the quality of life, extreme poverty and consumerism.

In this context, environmental education, according to the National Environmental Education Policy, is now considered to be:

> [...] an essential and permanent component of national education, and must be present, in an articulated way, at all levels and modalities of the educational process, in a formal and non-formal character (BRASIL, 1999).

According to Marcatto (2002), formal environmental education, which seeks to educate people on an ongoing basis, involves students in general, from nursery school to primary, secondary and higher education, as well as teachers and other professionals involved in courses and training. Informal education, on the other hand, involves all segments of the population, such as women's groups, youth groups, workers, politicians, businesspeople, residents' associations, liberal professionals, among others, and includes educational actions and practices aimed at raising awareness among the community about environmental issues and their organisation and participation in defending the quality of the environment.

However, according to Santos (2007), regardless of whether it is formal or informal, environmental education focuses on training environmentally committed citizens, whether they are of school age or not, with a critical sense, responsibility and participation in socio-environmental decisions.

2.2 Environmental problems

Since the emergence of life on Earth, human interventions have caused changes in ecosystems. Initially, according to Branco (1997), the environmental impact of anthropogenic action was little or almost non-existent. However, after the Industrial Revolution, according to Ferrari and Zancul (2008), the development of new technologies led to high levels of pollution, without concern for future generations.

According to Grun (2005), this view that nature is capable of providing inexhaustible resources, combined with human anthropocentrism, was the pivot of the ecological crisis, which is one of the elements responsible for the serious environmental problems faced today. These include the loss and extinction of animal biodiversity, air, water and soil pollution, whether caused by the disposal of sewage and rubbish or contamination with chemical pesticides and other products, deforestation and the burning of native vegetation to expand agricultural and livestock frontiers, soil impoverishment caused by incorrect use, and global warming and the depletion of the ozone layer due to gas emissions, all of which represent a major threat to life on the planet.

Authors such as Dias (1992) consider the rubbish produced in cities to be one of the biggest environmental problems, causing direct impacts on the local community. In this sense, the school, whose role is to train more responsible and aware citizens (ZEPPONE, 1999), must be directly linked to educational and environmental activities that promote a change in behaviour and awareness in these subjects, both in public and public schools.

However, in order for effective changes to take place, environmental education needs to be incorporated into educational processes, with a close link to the reality of these students (DIAS, 1992). In this sense, Kindel; Silva; Sammarco (2006) reported the need to involve and awaken a critical conscience in students, aiming to find solutions to the problem, which provides a perception of the world around them and the development of habits of co-operation, respect and responsibility within the family, educational and social environment.

As the world's population increases, so does the demand for food production. To achieve this increase in production, it is necessary to increase the area planted or increase production per unit of cultivated area. As a result, environmental impacts have grown at an accelerated pace, causing imbalances not just locally, but on a global scale. According to Branco (1997):

> [...] as the human species has developed new technologies and expanded its dominion over the elements and nature in general, environmental impacts have increased in intensity and extent (BRANCO, 1997).

In this sense, agricultural practiccs from the discovery of Brazil to the present day have been carried out, in most cases, inappropriately. The environment has been transformed by man, who destroys it and contributes, more often than not, to the extinction of animal and plant species on the planet, while also helping to pollute the air, soil and especially water through improper use. In short, man's predatory behaviour is not new, what is new is the proportion and extent of the mechanisms of depredation, which range from the emergence of monoculture plantations to nuclear weapons (VIOLA, 1987).

Pollution is:

> [...] an ecological alteration, i.e. an alteration in the relationship between living beings, caused by human beings, which harms, directly or indirectly, life or well-being, with damage to natural resources such as water and soil and impediments to economic activities such as fishing and agriculture (NASS, 2002).

Human activity has led to the pollution of rivers and lakes, and much of the water used to supply the population living in cities comes from these places. Every year, they become more polluted because they receive untreated domestic and industrial sewage, as well as chemicals used in agriculture, which are carried by rain from agricultural areas. This contamination of rivers, groundwater and springs reduces the availability of drinking water.

Another factor contributing to the decrease in the supply of drinking water is deforestation around springs, the silting up of rivers and soil sealing, which helps to reduce the replenishment of aquifers.

In cities, the problem of water pollution takes on catastrophic proportions, as they concentrate the largest populations and the majority of industries and services. There is high water consumption and a multitude of polluting sources, both in the form of domestic sewage and industrial effluents.

A large amount of water is also consumed by industrial activities, such as washing equipment and installations, cooling or heating processes, transporting industrial waste, manufacturing products in the chemical and beverage industries, among others. This huge amount of water used by domestic and industrial activities is not actually consumed, but taken from nature and then returned again, as it forms part of the hydrological cycle. That's exactly where the problem lies: most of this water, once used, is returned to the environment partially or totally polluted, i.e. it goes directly into streams, rivers, lakes and seas.

In this respect, we realise that water consumption in the world has been increasing in recent years and the availability of drinking water has been decreasing. Three quarters of our planet's surface is made up of oceans. However, the distribution of water reserves on the planet is uneven, and much of it is salty and found in the oceans and seas and cannot be used.

As a result, many places on the planet, such as cities and agricultural areas, are in serious danger of running out of water for good. The fundamental cause of this ecological tragedy, which could possibly get worse if serious measures are not taken, is the accelerated increase in various forms of pollution. Therefore, when considering that water is essential for the maintenance of life on Earth, one must be aware that predictions about future water scarcity require immediate action so that there is no shortage of treated water for generations to come.

Soil, in turn, is fundamental to the development of life and is an important component of terrestrial ecosystems, as it is the source of food products, but it is also directly affected by pollution. In this sense, soil pollution can be understood as any alteration caused to its characteristics by the introduction of chemical products or waste, in such a way that it becomes harmful to humans and other organisms, or has its uses jeopardised.

As it is a non-renewable resource, its use must be rational and planned. Inadequate and excessive use of the soil not only depletes it, but also reduces its natural fertility and organic matter. It also increases contamination through the use of chemical products, pesticides, domestic and industrial waste. Therefore, preserving soils, preventing them from being degraded, is a guarantee of biodiversity and arable land.

From this perspective, it can be inferred that human beings must realise that they are an integral part of nature and not superior to it. As Branco (1997) emphasises, man, regardless of his will, depends on the existence of a rich, complex and balanced nature around him. Thus, it is not enough to keep species alive, but in order to preserve the balance between human beings and nature, it is necessary to use natural resources consciously, prioritising development without affecting the quality of life of future generations.

In a complementary way, Philippi (2001) proposes sustainable development. According to this author,

> [...] In developing countries, the basic needs of a large number of people - food,
> clothing, housing, employment - are not being met. In addition to these basic needs, people legitimately aspire to a better quality of life. For there to be sustainable development, everyone's basic needs must be met and they must be given opportunities to realise their aspirations for a better life (PHILIPPI, 2001).

In this way, preserving the soil, preventing it from being degraded, is a guarantee of biodiversity and arable land. In addition, society's concern about the aggression of natural elements emphasises that activities for the development of humanity must be carried out in such a way as to preserve the essential qualities of natural resources, making it necessary to balance economic growth and the environment (CALGARO, 2012).

In recent decades, concern about environmental preservation has been growing worldwide, both on the part of civil society, governments and companies. More and more people are concerned about avoiding wasteful consumption, reducing rubbish production, saving water and energy, separating recyclable rubbish and reducing pollution, among other things. Together, these measures improve everyone's quality of life.

Following this line of thought, Pignatti (2004) stated that local environmental problems such as the degradation of air, water, soil, work and home environments have a significant impact on human health. In this sense, the aim is to solve the environmental problems caused by pollution so that they don't have repercussions for the population as a whole, seeking alternatives to alleviate these problems.

According to Currie (2000), when working on environmental issues, the teacher should emphasise the importance of the school for the community in which it is located and try to emphasise personal awareness aimed at particular responsibility for the environment, observation, organisation, analysis, communication, the use of imagination and creativity, promoting an integrated view of the world. In addition, according to the same author, the student's role in the community should be emphasised, establishing connections between their reality and school knowledge.

2.3 The role of the school and the family in environmental education

The environmental problems faced by society are increasingly evident and require environmental education actions that are effective in promoting environmental awareness among citizens. In order to promote this behavioural change, Guimarães (2005) reported that it is necessary to implement environmental education aimed at the population in general, but mainly at the younger generations, who are at an age when values and attitudes are being formed.

Environmental education in the school environment is guaranteed by Law No. 9,795 of 27 April 1999. According to this legislation, education must:

> [...] be developed within the curricula of public and private educational institutions, encompassing basic education (early childhood education, primary education and secondary education), higher education, special education, professional education and youth and adult education". In addition, "environmental education will be developed as an integrated, continuous and permanent educational practice at all levels and modalities of formal education, and should not be implemented as a specific subject in the teaching curriculum (BRASIL, 1999).

In this sense, the school, defined as an institution geared towards teaching students, emerges as a link between environmental education and society, and can effectively contribute to the development of each individual's environmental sense . Its function is to prepare and train individuals for life in society (ALARCÃO, 2001). In order for this to be possible, according to Moradillo and Oki (2004), it is necessary to articulate theoretical and practical knowledge, so that teachings take on meaning in each individual's life:

> The role of the school, from a non-naive political perspective, is to create spaces through its actors and social authors in order to de-alienate individuals from fragmented knowledge that has no meaning for their social actions (MORADILLO; OKI, 2004).

One of the school's roles is to help build values and strategies that enable students to determine the best way to maintain and improve their cultural, natural and economic heritage in order to achieve sustainability (TRISTÃO, 2008). Therefore,

> The role of the school as an agent of education is fundamental in developing a healthy relationship that brings benefits to it as an institution, to all its human components, but also to the environment as a whole, both local and global (RIBES, 2000).

In this sense, according to Reigota (1998), environmental education activities developed in the school environment should focus on raising awareness, changing behaviour, developing skills, the ability to evaluate and the participation of students.

In addition, it is necessary to overcome the fragmentation of knowledge through an objective interdisciplinary approach that enables the development of environmental education at school. Corroborating this statement, Reigota (1998) emphasises that school environmental education was founded on the perspective of transmission or construction of knowledge, which allows it to be developed from different aspects that complement each other.

True environmental education must provide citizens with a basis for self-criticism, favouring the understanding that man is also nature and must therefore respect his limits and those of humanity (COSTA, 2008). In this sense, environmental, economic, political and social challenges are interconnected and require a view of the whole and not just the parts.

According to Pádua and Tabanez (1998), environmental education encourages an increase in knowledge, a change in values and the improvement of skills, which are basic conditions for encouraging greater integration and harmony between individuals and the environment. Faced with the complexity and magnitude of man's relationship with the environment, environmental education takes on a challenging role that requires new knowledge on the part of educators and students in order to develop a critical and transformative environmental education that promotes the exercise of effective citizenship.

However, it is not only the school that has the task of promoting environmental education. According to Bonachela and Marta (2010), environmental education should take place both inside and outside educational establishments, and is not restricted to school education, which takes place only in educational establishments.

The family, regardless of the family arrangement, is the indispensable place for guaranteeing the survival and integral protection of children and other members. It also provides the emotional and material support necessary for the development and well-being of its members (KALOUSTIAN, 1988). According to Bonachela and Marta (2010):

> [...] It plays a decisive role in formal and informal education; it is where ethical and humanitarian values are absorbed and where bonds of solidarity are deepened. It is also where the marks between generations are built and cultural values are observed (BONACHELA; MARTA, 2010).

According to Soifer (1982), it is in the family, the basic social structure, where the first relationships take place with a view to forming a group, providing the foundations for future experiences. Thus, for Pires and Brombergersim (2007):

> [...] the family has a great influence on the formation of the individual, whether socially, educationally or culturally (PIRES; BROMBERGERSIM, 2007).

By guaranteeing everyone the right to an ecologically balanced environment, which is a common good and essential to a healthy quality of life, the 1988 Federal Constitution assigns the public authorities and the community the duty to defend and protect the environment for present and future generations. In this sense, Bonachela and Marta (2010) characterise the community as a group or system of social relationships and recurring interactions between people, in which individuals organise themselves into even smaller groups called families, which play a significant role in the informal education of the individual.

It should be emphasised that education generally takes place through the observation and interaction of individuals, which enables the transmission and assimilation of information. That's why, in order to promote effective environmental education, it's necessary to involve all players, whether they're at school, in the family or in society.

2.4 Rural and urban schools and their involvement in environmental education

With the creation of the Law of Guidelines and Bases for Education (LDB), Law No. 9.394/96, sociocultural diversity was recognised, and it was up to education systems to make adjustments to the peculiarities of rural and/or urban life in each region. For Zakrzevski (2007), these adjustments in rural schools should involve curricular content and methodologies appropriate to the needs and realities of the students, adapting the school calendar to the agricultural phases and climatic conditions, among others, and not proposing a simple and pure adaptation of urban education to the rural environment.

In this sense, Caldart (2003) stated that the rural school has a real meaning. For him:

> The rural school is not a different kind of school, but rather a school that recognises and helps to strengthen rural people as social subjects who can also help in the process of humanising society as a whole, with their struggles, their history, their work, their knowledge, their culture, their way (CALDART, 2003).

According to Furtado (2004), rural schools have their own peculiarities due to the dispersion of the resident population, and are usually small in terms of the number of students they serve. Nevertheless, according to the same author, these schools, with their physical and human limitations, are marked by high dropout and repetition rates and high teacher turnover. However, even with these limitations,

> [...] it is necessary to take into account that, even today, it is only this weak institution that carries the considerable task of educating a large number of Brazilians of all ages spread across the countryside of this country of continental dimensions (SOARES, 2007).

Given these facts, in order to ensure their effectiveness, rural schools need to implement specific environmental education. According to Zakrzevski (2007), this should be based on a specific context, geared towards the interests and needs of the people who live and work in the countryside, linked to the demands and specificities of each locality or community. Thus, according to this author, the biggest challenge for environmental education in rural schools is to stimulate a process of reflection on responsible, economically viable and socially acceptable models of rural development that help to reduce poverty, conserve natural resources and biodiversity, transform socio-environmental problems and strengthen communities, without dissociating the complexity of society and nature.

With regard to the urban environment, urbanisation represents the development of cities and is accompanied by population growth, which requires more and more infrastructure in cities (LIMA, 2009). According to this author, when all this development is linked to planning, urbanisation brings benefits to society and the environment. Otherwise, however, social and environmental problems multiply in cities.

In cities where this occurs, conflicts between man and nature are evident. In this context, environmental education in urban schools should be seen as an instrument for implementing sustainable development, combining economic and technological growth with the rational exploitation of natural resources and their conservation for future generations (BRITO; CASTRO, 2003). Environmental education in these places and for this public must develop a new teaching model, aimed at promoting new attitudes, attitudes and values compatible with a culture of sustainability.

CHAPTER 3

MATERIAL AND METHODS

3.1 Location and target audience

The Érico Veríssimo State High School, where the research was carried out, is located at Rua Izaltino de Oliveira, 164, in the centre of Restinga Sêca/RS. The school has five hundred and ninety students enrolled in polytechnic, regular and youth and adult education, in the morning, afternoon and evening shifts, catering for students from both rural and urban areas. The age range of the students is predominantly between fifteen and nineteen, due to the different types of education offered by the school. The questionnaire was administered to 11 secondary school classes. Due to the researcher's availability, the questionnaires were administered to a larger number of students on the morning shift, as she has a heavier workload on this shift. Likewise, we would like to point out that most of the students from rural areas study during this shift, which may be important for interpreting the results.

3.2 Assessment tools

In order to carry out this research, a questionnaire containing thirteen closed questions was administered to 311 students in the 1st and 2nd year of secondary school, polytechnic mode, and students in the 3rd year of the old secondary school, i.e. 50% of the students regularly enrolled at the school responded to the questionnaire (Appendix 1). Despite the closed questions, the students were free to express their opinion, based on comments they felt were pertinent at the time, either in writing, attached to the questionnaire, or in class, after it had been applied. The questions were applied during regular Geography lessons, over a period of one week, in the grades mentioned above.

CHAPTER 4

RESULTS AND DISCUSSION

A total of 311 questionnaires were answered by all the classes evaluated. Analysing the data, it was possible to see that 60% of the students live in rural areas (Figure 1). It is important to emphasise here that most of the questionnaires were administered during the morning shift, where young people from rural areas predominate, a fact that should be taken into account in subsequent inferences.

Most of the interviewees' families are relatively small (Figure 2), with only 12 per cent of the students coming from large families.

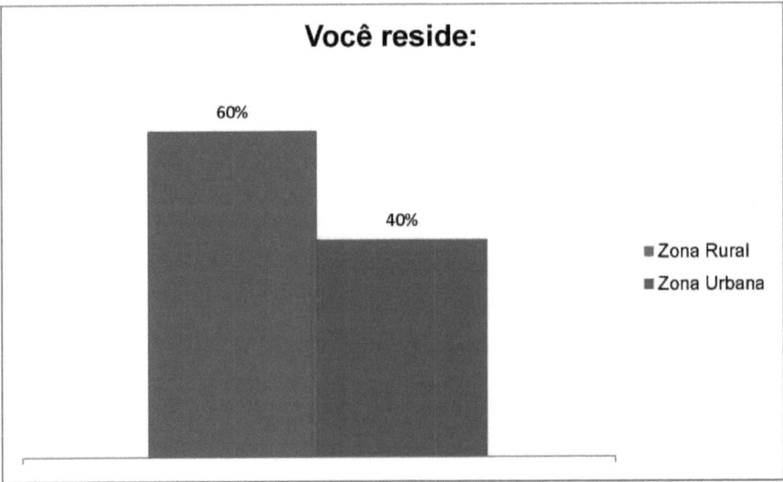

Figure 1 - Origin of the students interviewed: rural or urban area. E. E. M. Érico Veríssimo (Restinga Sêca, RS, Brazil).

This result is in line with data from the IBGE (2009), which showed that families are becoming smaller and smaller as a result of couples opting for fewer children. However, it should be borne in mind that the result does not show the number of children in each family, but rather the number of people living in each house.

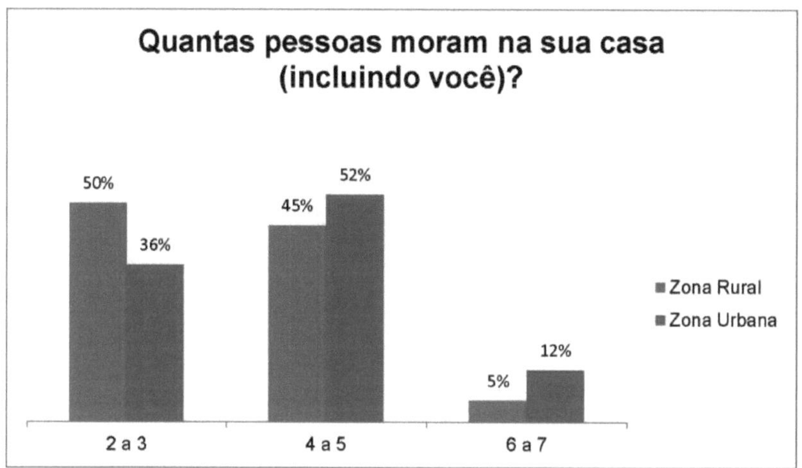

Figure 2 - Number of family members of students at E. E. M. Érico Veríssimo (Restinga Sêca, RS, Brazil). .

Analysing Figure 2, it can be inferred that urban families are more numerous, but there may be a link to the rural exodus. According to data from the IBGE demographic census (2010), the number of people living in rural areas continues to fall in the country, but at a slower rate than in the previous decade. The country's rural population lost two million people between 2000 and 2010, which represents half of the four million who moved to the cities in the previous decade.

Questions three to thirteen, analysed below, are the most directly related to the issue of household waste.

Figure 3 shows that in the vast majority of households there is regular rubbish collection by the competent municipal bodies, even in rural areas. However, the figure shows that some families don't consider regular collection to be monthly. This requires rural dwellers to separate, burn and consider another destination for their rubbish to avoid it being left exposed until the next collection.

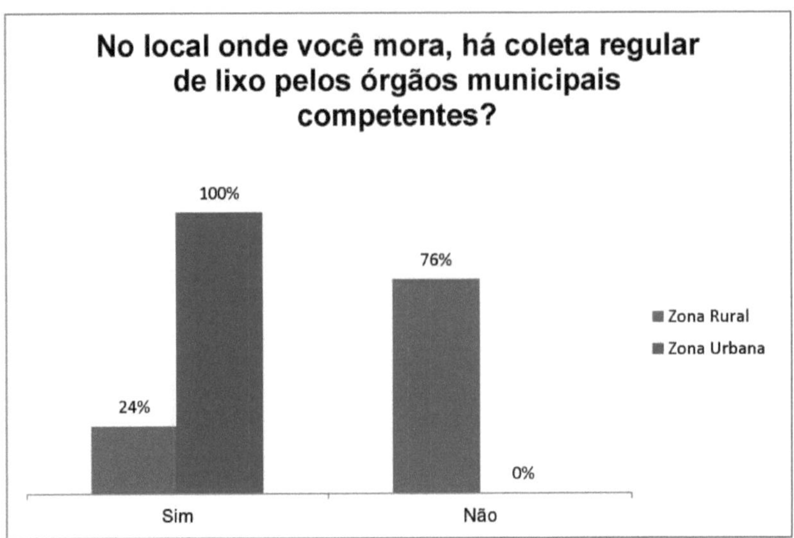

Figure 3 - Characteristics of the regularity of household rubbish collection for families of students at the Érico Veríssimo school (Restinga Sêca, RS, Brazil).

Data from the Brazilian Institute of Geography and Statistics (IBGE, 2003) revealed that rubbish collection in rural communities reaches only 20 per cent of households. As a result, waste is often disposed of incorrectly in nature.

However, these substances contaminate the soil and can reach water sources, contaminating other areas or parts of the land (BRASIL, 2005). Rubbish placed in an inappropriate place can not only degrade the landscape and produce a bad odour, it can also put public health at risk. According to this author, as it provides abundant food, it can attract insects, dogs, rats and other animals, which can directly or indirectly spread dozens of diseases. As well as causing soil, water and air pollution.

According to Rouquayrol (1994), rural areas also need attention and correct sanitation solutions, as this will prevent contamination of the environment and possible diseases that could affect human beings as a result of poor waste disposal. However, in order for society to rethink its habits, it is important that environmental education and awareness work together to achieve the necessary changes (BRASIL, 2008). In this sense, public policies are needed to alleviate the issue of rubbish and, consequently, the environmental degradation it causes. It is worth mentioning that, in an attempt to alleviate the environmental problems caused by the improper disposal of rubbish, the Federal Government passed Law 11.445/07, which establishes national guidelines for basic sanitation and the National Basic Sanitation Policy. However, for the families analysed, there is no selective collection of household waste (Figure 4).

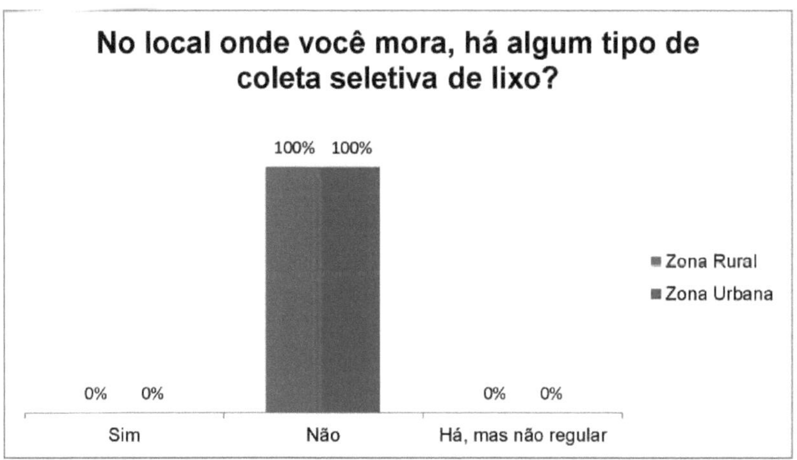

Figure 4 - Characteristics of the selectivity of household rubbish collection in the homes of students at the Érico Veríssimo school (Restinga Sêca, RS, Brazil).

This highlights the lack of public policies on the subject, which could result in a major environmental impact in the long term. Data from the IBGE (2011) showed that 67.7 per cent of municipalities do not have selective waste collection.

In the urban area, the vast majority of families don't separate their rubbish, i.e. 47%; 28% of families do and 28% separate it eventually (Figure 5). In rural areas, the result is more uniform, i.e. 35% separate their rubbish, 35% say they don't and 25% do it eventually.

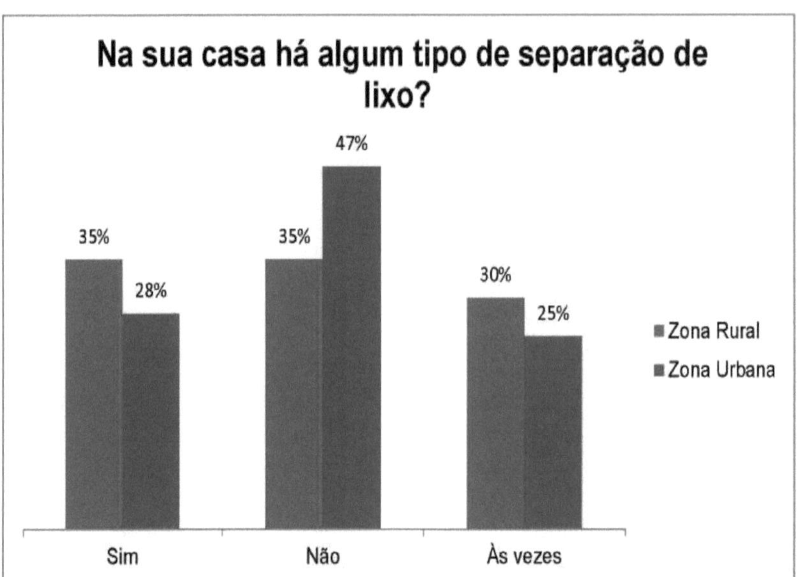

Figure 5 - Household waste separation in the families of students at the Érico Veríssimo school (Restinga Sêca, RS, Brazil).

This figure shows that the practice of separating rubbish is more consistent in rural areas, which may be considered an unexpected result at first. The students' comments after the questionnaires showed that, in fact, rural families separate rubbish because they make better use of it; with organic rubbish they feed the animals and make soap, for example. With regard to dry rubbish, the students said that this type of waste is burnt in their homes. These facts can also be seen in Figure 6.

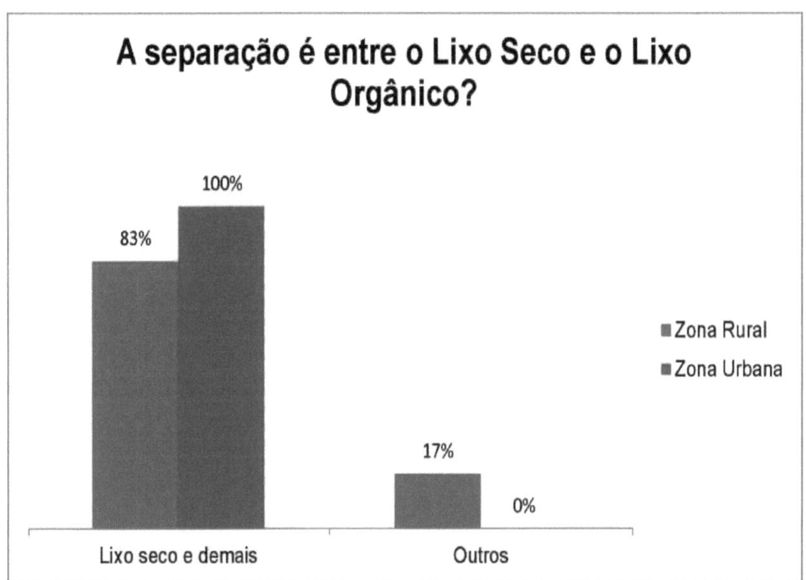

Figure 6 - Household rubbish sorting in the families of pupils at the Érico Veríssimo school (Restinga Sêca, RS, Brazil).

Figure 7 shows how unconcerned families are about consuming natural or durable products, since the answers were the same in both rural and urban areas, in terms of being careful about buying products that could generate some kind of waste when disposed of.

In rural areas, 77 per cent of families reported that they did not worry about the waste they would generate when buying products. Only 23% reported this concern when purchasing the products. In urban areas, this concern, or lack of it, is also present, as 79 per cent of families reported not using reusable products and only 21 per cent were concerned about the amount of waste that will be disposed of in the environment.

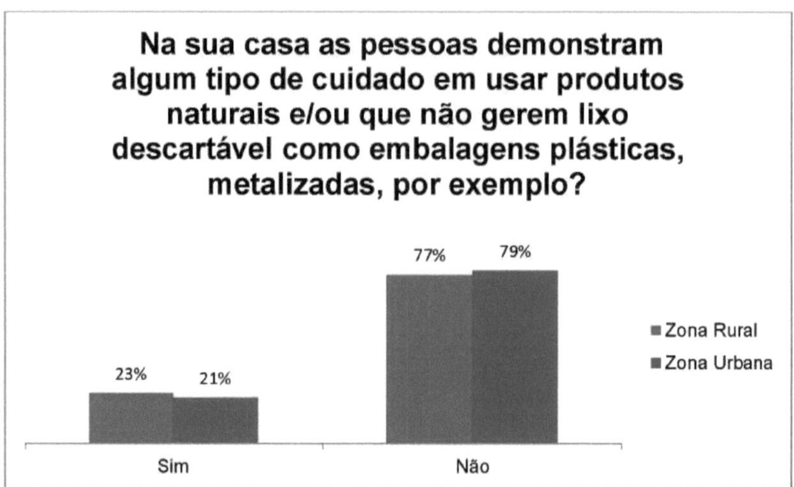

Figura 7 Environmental awareness regarding the use of disposable products among families of students at the Érico Veríssimo School of Education (Restinga Sêca, RS, Brazil).

The vast majority of people don't mind consuming products that generate disposable waste. Perhaps this is due to the rampant industrialisation of food. This issue could have stood out among urban families, who have more access to this type of product, but the answer is in line with what was found in rural families.

According to Portal Brasil, the average amount of rubbish produced daily per inhabitant in our country is 1kg (BRASIL, 2013). The figure seems small, but it's huge when the IBGE points out that we already have over 190 million inhabitants (IBGE, 2010).

When the students were asked how they disposed of used cooking oil at home (question 8), there was a level playing field between the answers (Figure 8). In rural areas, 18% of people dispose of used oil in the kitchen sink; 5% are concerned about the environment and store it for later disposal in accredited places and 77% of this population, the vast majority, use the oil for other purposes, i.e. preparing food for animals and making soap.

In the urban area, 23 per cent dispose of it in the kitchen sink; the majority, 40 per cent, dispose of it directly on the ground; 16 per cent store it for later disposal at accredited sites, i.e. the minority, and 21 per cent dispose of it in other places, i.e. store it for later soap making, as described by the students in their observations on the questionnaires. These results show an interesting question: if there is no selective or even regular rubbish collection in rural areas (Figures 3 and 4), it may seem discrepant that 16 per cent of rural families store cooking oil for later disposal (Figure 5).

It's likely that this result is a manifestation of embarrassment, which already indicates an awareness of the problem. However, a considerable proportion of families, both rural and urban, dispose of waste down the sink. This uncompromising attitude causes sewage pipes to clog, increasing treatment costs by up to 45 per cent (BIODIESEL, 2008). Also noteworthy is the large proportion of families in rural areas who dispose of cooking oil on the ground (40 per cent).

Figure 8 - Where cooking oil is disposed of by the families of students at the Érico Veríssimo school (Restinga Sêca, RS, Brazil).

Although research has already shown that one litre of cooking oil that ends up in the water contaminates around one million litres of water, equivalent to the consumption of one person in 14 years, only now do environmentalists agree that there is no ideal disposal model for the product, but rather alternatives for reusing frying oil to make biodiesel, soap, among other things (AMBIENTE EM FOCO, 2008). This shows that this type of pollution can lead to ecological imbalance, threatening the lives of various animal and plant species.

Figure 9 shows that 71% of families in rural areas are in the habit of disposing of expired medicines in ordinary rubbish and 29% reported that they are in the habit of keeping them at home and then disposing of them at accredited sites when travelling to urban areas. However, it should be emphasised that this information may not be entirely in line with reality. After the questionnaire was administered, it was clear from talking to the students that they were embarrassed to report their family's habit of disposing of medicines, just as they had been about disposing of cooking oil. If, on the one hand, this attitude masks the intended results, on the other it shows that the students have understood the seriousness of people's behaviour in this area. It is very likely that the majority had never considered this attitude; it is hoped that this issue will be brought up in the family, which could be the start of a change in behaviour.

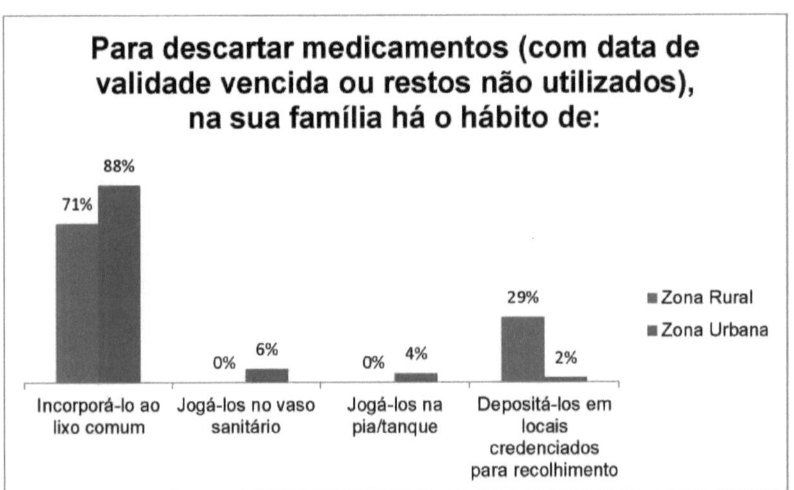

Figure 9 - Disposal of medicines by families of students at the Érico Veríssimo school (Restinga Sêca, RS, Brazil).

When comparing the responses of families in urban areas regarding the disposal of medicines at accredited sites, it is clear that only a minority (2%) claim to properly dispose of medicines, despite the fact that they have greater availability and ease of finding these sites, since they live in urban areas; 88% of the students reported that the family incorporates expired medicines into the common rubbish; 6% throw them straight into the toilet and 4% into the kitchen sink.

Estimates have shown that the Brazilian population currently generates around 10,300 tonnes a year of medicine waste without a proper disposal system (ANVISA, 2013). The National Health Surveillance Agency (ANVISA) warns consumers about the risks of disposing of medicines incorrectly. Many people, for lack of alternatives and information, still throw expired medicines or medicines that will no longer be used in the rubbish or sewage system. According to ANVISA, this practice can contaminate water and soil (ANVISA, 2013).

Today, a large proportion of medicines go into the rubbish bin or sewage system (when flushed down the toilet); as the use of medicines increases, so does this improper disposal, which possibly leads to greater contamination of groundwater, rivers and other negative consequences (SAUDEWEB, 2013).

In this vein, Figure 10 shows that 12% of families in rural areas buy rubbish bags to store their rubbish; 41% use shopping bags from the market for this purpose and 47% burn their rubbish, since, as shown in Figure 3, families in rural areas do not have regular rubbish collection and therefore need to dispose of their rubbish so that it does not become a breeding ground for mosquitoes or others.

In the urban area, 16% of families buy suitable bags to dispose of their rubbish and 84% reuse shopping bags from the market, meaning that the vast majority of the population living in urban areas have this habit. This is a serious problem, as a plastic bag takes between 100 and 400 years to decompose.

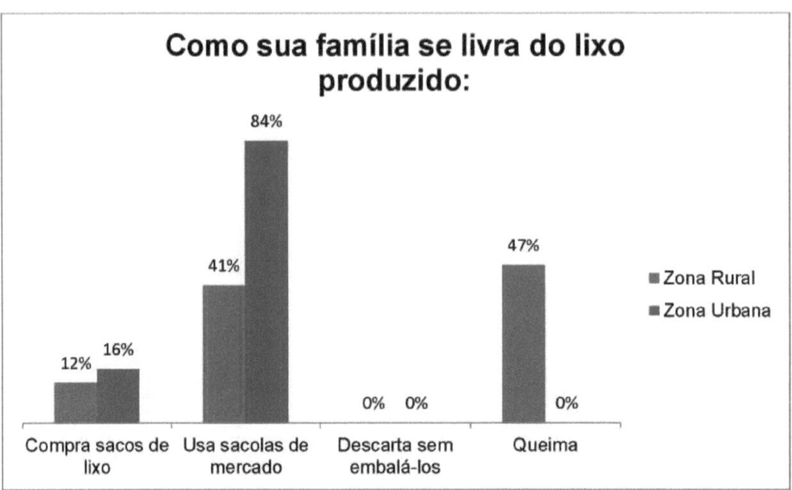

Figure 10 - Household rubbish disposal by the families of students at the Érico Veríssimo school (Restinga Sêca, RS, Brazil).

With regard to this packaging, it is interesting to know that

> supermarket rubbish bags are specifically designed for this purpose. They can be made from a mixture of recycled material, using other types of plastic and even other bags. Supermarket bags, on the other hand, are not, because as they will receive food, they usually have to be made from 100% virgin raw materials. This in itself makes rubbish bags more suitable and less harmful (REVISTA CRESCER, 2013).

Data from the IBGE (2010) showed that in more than half of Brazilian households in rural areas, the lack of a waste collection system leads residents to a dangerous practice: burning rubbish. The difficulty and high cost of collecting rural rubbish makes the option of burning it the one most adopted by residents in these regions (Figure 10).

With regard to smoking and waste disposal, it can be seen that in rural areas 35% of people smoke and 65% do not (Figure 11). In the urban area, the percentage of smokers is 31% and 69% non-smokers. This shows an equalisation in both places of origin of the students surveyed and the most interesting thing is that the percentage of non-smokers has been increasing over the years.

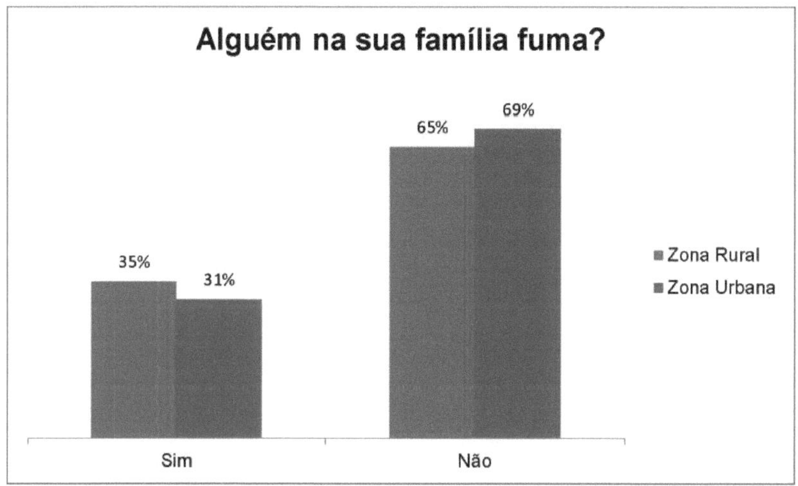

Figure 11 - Existence of smokers in the families of students at the Érico Veríssimo secondary school (Restinga Sêca, RS, Brazil).

Corroborating the results shown in Figure 11, research carried out by the National Cancer Institute (INCA) showed that the number of Brazilians who smoke regularly is falling (INCA, 2013). The smoking rate in the country has fallen by almost 50 per cent in the last 25 years.

According to the survey, if Brazil had not implemented any tobacco control measures, the percentage of smokers in 2010 would have been 31 per cent. This means that approximately one in three people aged 18 or over would be smokers.

The success in reducing tobacco consumption is mainly due to the laws that restricted advertising and banned smoking in enclosed spaces. The Anti-Smoking Law (No. 13,541/09) has helped to reduce the number of smokers by 9 per cent, according to data released by INCA.

With regard to the destination of cigarette butts (Figure 12), we can see that in rural areas, 17% of smokers in their family throw them directly in the yard or in the street; the same percentage of 17% dispose of cigarette butts in rubbish bins. In addition, a large percentage of those interviewed replied that their family members dispose of it without any criteria, i.e. 50% throw it anywhere and 16% replied that they don't know where it is deposited.

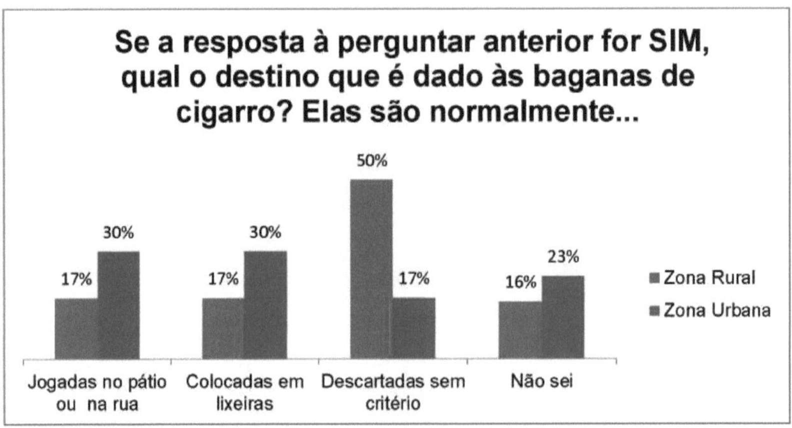

Figure 12 - Destination of cigarette butts, practice of smokers in families of students at the Érico Veríssimo School of Education (Restinga Sêca, RS, Brazil).

In urban areas, 30 per cent throw cigarette butts in the yard or on the street; the same percentage, 30 per cent, put them in rubbish bins; 17 per cent dispose of them without any criteria and 23 per cent answered that they don't know where they go.

Thus, the incorrect disposal of cigarette butts harms rainwater runoff and studies indicate that 25 per cent of fires in urban areas are caused by cigarette butts (ECOCITYBRASIL, 2013). According to the same author, even in cases where establishments collect them, they are disposed of together with ordinary rubbish, thus polluting the groundwater due to the chemical components contained in the waste. The oceans also suffer from cigarettes, as packaging and cigarette butts accumulate in the seas, making up 40 per cent of the rubbish found.

As an interesting example of how to dispose of this type of waste, the company Ecocity, located in Curitiba, PR, created the Bituca Zero Programme, installing cigarette butt collectors, which are transformed into cellulose acetate blankets and used for hydro-seeding, i.e. the tobacco and paper become fertiliser that can later be applied to degraded areas (ECOCITYBRASIL, 2013).

Finally, the last figure (Figure 13) shows a satisfactory result in terms of the families' environmental awareness regarding the destination of the rubbish produced and the problems that improper disposal can cause for people's health.

When the students were asked if they believed that the correct collection and storage of rubbish could make a difference to the health of the environment, there was an almost unanimous positive response, especially among students from rural areas, i.e. 100% said yes, that they were aware of the environmental issue. As for the students from urban areas, 93% understood the need to raise awareness and care for the environment and only 7% reported not being concerned about this issue. However, the concern, especially with those who responded negatively to the concern about environmental problems, is evident, as these students need differentiated awareness-raising and intervention work to resolve the environmental issue.

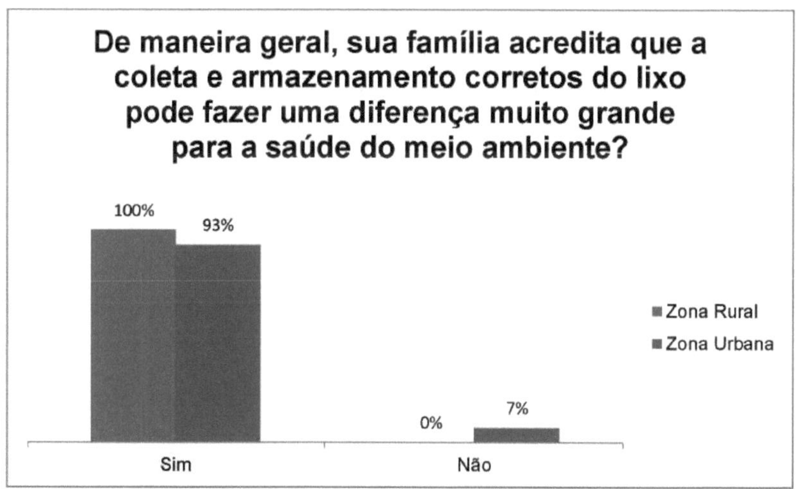

Figure 13 - Awareness of the families of students at the Érico Veríssimo school (Restinga Sêca, RS, Brazil) regarding the collection and disposal of household rubbish.

The results show that these students and their families are very sensitive to the environmental issue linked to rubbish; this demonstrates that they are aware of the importance that the process of sorting household rubbish has for the population in general .

According to the IBGE, separating and recycling paper, glass, plastic and metal, for example, removes a series of materials from the rubbish that would take an astonishing amount of time to decompose - such as plastic (450 years), aluminium cans (200 years) or glass (1 million years). What's more, by being reused, recyclable waste saves natural resources. Elen Aquino, a biologist and researcher at the Centre for Training and Research in the Environment (Cepema) at the University of São Paulo, said that:

> One tonne of recycled paper saves 22 trees, 75% of electricity and pollutes the air 74% less than producing the same amount of paper with virgin raw materials (VEJABRASIL, 2013).

The production, separation, collection and disposal of rubbish is a matter of responsibility and citizenship, and therefore everyone's duty.

CHAPTER 5

CONCLUSION

By analysing the results obtained, it can be inferred that both rural and urban families are aware, in theory, of the correlation between inadequate treatment of household waste and environmental damage. However, practice does not seem to confirm this premise. Families, regardless of their region of origin, don't know or don't value minimum precautions such as the proper disposal of cooking oil or the use of products that generate little waste. While families in rural areas prioritise burning their household waste, those in urban areas make massive use of plastic bags to dispose of it; while the latter pollute the soil with cooking oil, half of rural families don't have any criteria for disposing of cigarette butts.

In this way, it seems correct to say that although families have already been sensitised, there is still a long way to go when it comes to caring for the environment. Environmental education is currently the most effective tool for creating and implementing sustainable forms of interaction between society and nature. This is the way for each individual to change their habits and adopt new attitudes that will lead to a reduction in environmental degradation, improve quality of life and reduce the pressure on environmental resources.

CHAPTER 6

FINAL CONSIDERATIONS

Concern about rubbish is one of the environmental problems that threaten life on Earth, because as well as polluting the soil, water and air, it also attracts countless insects and animals that cause disease. Concern about the environment should be part of the life of every citizen and their leaders. Everyone should make the place where we live a pleasant and healthy place. It is from this perspective that environmental education has become a concern for a large part of humanity.

In order for socio-environmental change to take place, awareness and reflection promoted by Environmental Education must be paramount, as well as society taking action by learning about citizenship. In this sense, Environmental Education seeks a balance between man and the environment. In order to build a sustainable future, there needs to be a transformation in human behaviour in relation to the environment. This study emphasises the issue of disposing of household waste, and this diagnosis could subsidise future actions within the school environment aimed at remedying any shortcomings.The research found that the population uses different procedures to dispose of household waste. In this way, it can be seen that the collection carried out by the municipal government in rural areas is still very precarious. Ideally, collection should take place on a regular basis and not just once a month, as the students reported throughout the survey. Another alternative would be to build a place where rural dwellers could store their rubbish so that it could later be collected by the competent body.

BIBLIOGRAPHICAL REFERENCES

NATIONAL HEALTH SURVEILLANCE AGENCY - ANVISA. Available at: < www.anvisa.com.br> Accessed on 22 November 2013.

ALARCÃO, I. **Escola reflexiva e nova racionalidade.** Porto Alegre: Artmed, 2001, 82p.

AMBIENTE EM FOCO. **Recycling cooking oil can help reduce global warming.** 2008 Available at: <www.ambienteemfoco.com.br> Accessed in October 2013.

BARRETO, V. P. **Environmental Education as a proposal that reflects reality. Applied general studies centres**. Pedagogy Course Monograph. UFF, 2006. 75p.

BIODIESEL. **Recycling cooking oil.** 2008. Available at: <www.biodieselbr.com>Accessed October 2013.

BONACHELA, D. P.; MARTA, T. N. Environmental education: an important role for the family. **Revista de Direito Público**, Londrina, v. 5, n. 3, p. 236-253, dec. 2010.

BRAGA, A. R. A **influência do Projeto "A formação do professor e a Educação Ambiental" no conhecimento, valores, atitudes e crenças nos alunos no Ensino Fundamental**. Master's dissertation. Faculty of Education, State University of Campinas. Campinas, SP: 2003. 243p.

BRANCO, S. M. **O Meio Ambiente em Debate**. Ed. 26, São Paulo: Moderna. 1997. 87 p.

BRAZIL. Ministry of the Environment. **Education manual for sustainable consumption**. Brasília: MMA, 2005.

______ . Ministry of the Environment. The different shades of Environmental Education in Brazil. 1997 - 2007. Brasília: MMA, 2008.

______ . Ministry of the Environment. **National Environmental Education Policy.** Brasília, 1999. Available at: http:// http://www.planalto.gov.br/ccivil_03/leis/l9795.htm. Accessed: 10 August 2013.

BRITO, H. O. E.; CASTRO, C. **Environmental education and sustainable development.** Natal: Davinci UFRN, 2003.

CALDART, R.. The rural school on the move. **Currículo Sem Fronteiras**, v. 3, n. 1, p. 60-81, jan./jun. 2003.

CALGARO, C. Article on Sustainable Development: a reality to be achieved. 2012, **Revista Jus Vigilantibus on.** Available at: <http:// http://repositorio.ub.edu.ar:8080/xmlui/bitstream/handle/123456789/1479/mes a3-5.pdf?sequence=1>. Accessed on 09 January 2013.

CARVALHO, I. C. M. **Educação Ambiental:** Formação do Sujeito Ecológico. 2° ed. São Paulo Cortez, 2006.

COSTA, F. A. M.. **Socio-environmental educommunication**: popular communication and education. Brasília: MMA, 2008.

CURRIE, K. **Meio Ambiente:** Interdisciplinaridade na prática. Campinas-SP, Papirus, 2000.

DESSEN, M. A.; POLONIA, A. C. **Family and school as contexts for human**

development. Paidéia (Ribeirão Preto), Ribeirão Preto, v. 17, n. 36, Apr. 2007. Available in: <http://www.scielo.br/scielo.php?script=sci_arttext&pid=S0103-863X2007000100003&lng=en&nrm=iso>. Accessed on 24 July 2013>.

DIAS, G. F. **Educação Ambiental:** Princípios e Práticas. 3ª ed. São Paulo: Gaia, 1992.

DONELLA, M. **Conceitos para se fazer Educação Ambiental**. Department of the Environment, 1997.

ECOCYTYBRASIL ENVIRONMENTAL SOLUTIONS. Available at: <http://www.ecocitybrasil.com.br/portal/servicos/bituca-zero.html>. Accessed on 21 November 2013.

FERRARI, A. H.; ZANCUL, M. C. S. Environmental education: from the political-pedagogical project to the classroom. **Educação em Revista**, Marília, v.9, n.1, p.19-34, jan.-jun. 2008.

FURTADO, E. D. P. **The state of the art of rural education in Brazil.** Fortaleza: FAO/UNESCO, 2004.

GUIMARÃES, M. **A dimensão Ambiental na educação.** Campinas-SP: Papirus, 2005.

GRUN, M. The concept of holism in environmental ethics and environmental education. In: SATO, M.; CARVALHO, I. C. M. **Educação ambiental:** pesquisa e desafios. Porto Alegre: Artmed, 2005. p. 45-50.

BRAZILIAN INSTITUTE OF GEOGRAPHY AND STATISTICS - IBGE. Available at: http://ibge.gov.br<Accessed 17 October 2013>.

BRAZILIAN INSTITUTE OF GEOGRAPHY AND STATISTICS - IBGE. Available in: <http://www.ibge.gov.br/home/estatistica/populacao/trabalhoerendimento/pnad2003/> Accessed 20 November 2013.

BRAZILIAN INSTITUTE OF GEOGRAPHY AND STATISTICS - IBGE. Available in:

http://www.ibge.gov.br/home/estatistica/economia/perfilmunic/2009/ Accessed on 21

November 2013. BRAZILIAN INSTITUTE OF GEOGRAPHY AND STATISTICS -

IBGE.

Available in: <http://www.ibge.gov.br/home/estatistica/populacao/censo2010/ Accessed on 21

November 2013. BRAZILIAN INSTITUTE OF GEOGRAPHY AND STATISTICS -

IBGE.

Available in: http://www.ibge.gov.br/home/estatistica/populacao/estimativa2011/estimativa. shtm

Accessed on 22 November 2013. BRAZILIAN INSTITUTE OF OPINION AND

STATISTICS - IBOPE.

Available at:< http://ibope.com.br> Accessed on 17 October 2013.

NATIONAL CANCER INSTITUTE - INCA. Available at: <http://www.inca.gov.br/tabagismo> Accessed on 20 November 2013.

KALOUSTIAN, S. M. (org.) **Brazilian family, the basis of everything.** Brasília: UNICEF, 1988.

KINDEL, E. A. I.; SILVA, F. W. S.; SAMMARCO, Y. M. **Educação Ambiental:** Vários Olhares e Várias Práticas. 2ª ed. Curitiba, PR. Mediação, 2006.

LIMA, E. **Environmental education and urban space:** a reflection on nature-society. University of Paraíba, 2009.

MARCATTO, C. **Environmental education:** concepts and principles. Belo Horizonte: FEAM, 2002. 64 p.

MEDINA, N. M. Training multipliers for Environmental Education. In: PEDRINI, A. G. (org.). **O contrato social da ciência.** Petrópolis: Vozes, 2000. p.69-90.

MORADILLO, E. F.; OKI, M. C. M. Environmental Education at the University: Building possibilities. **Química Nova,** v. 27, n. 2, p. 332-336, 2004.

NASS, D. P. **The concept of pollution. Revista Eletrónica de Ciências -** n.

13, November 2002. Available at: http://www.cdcc.usp.br/ciencia/artigos/art_13/poluicao.html. Accessed: 09 August 2013.

PÁDUA, S.; TABANEZ, M. (eds.). **Environmental education:** paths travelled in Brazil. São Paulo: Ipê, 1998.

PHILIPP, L. S. The Construction of Sustainable Development. In: LEITE, A. L. T. A; MININNI-MEDINA, N. **Educação Ambiental (Curso Básico à Distância) Environmental Issues - Concepts, History, Problems and Alternatives**. 2. ed. v. 5. Brasília. Ministry of the Environment, 2001.

PIGNATTI, M. G. **Saúde** e Ambiente: Emerging diseases in Brazil.
Revista Ambiente & Sociedade - vol. VII n°.1 jan - jun, 2004.

PIRES, P. A. G.; BROMBERGER, S. M. T. A educação ambiental e o trabalho de cidadania com adolescentes. **Revista Ambiente & Educação**, vol. 12, 2007.

REIGOTA, M. Challenges to school environmental education. In: JACOBI, P. et al. (eds.). **Education, environment and citizenship:** reflections and experiences. São Paulo: SMA, 1998. p.43-50.
REVIEW CRESCER, 2013. Available at: < http://revistacrescer.globo.com/Revista/Crescer/0,,EMI203353-17334,00-SACOLAS+PLASTICAS+X+SACOS+DE+LIXO+QUAL+E+A+MELHOR+OPCAO.html> Accessed 21 November 2013.

RIBES, E. L. School and the environment - a productive exchange. In: LAMPERT, E. **Educação Brasileira:**desafios e perspectivas para o século XXI. Porto Alegre: Sulina, 2000, p. 75-87.

ROUQUAYROL, Maria Zélia (Org.). Epidemiologia & Saúde. 4 ed. Rio de Janeiro: MEDSI, 1994.

RUSCHEINSKY, A. **Sustentabilidade: uma paixão em movimento**. Porto Alegre: Sulina, 2004.

SANTOS, B. S. **Para uma revolução democrática da Justiça.** São Paulo: Cortez, 2007.

SATO, M. **Educação Ambiental**. São Carlos: Rima, 2004.

SAUDEWEB. Available at <http://saudeweb.com.br/39885/brasileiros- descartam-103-toneladas-de-remedios-por-anno/>. Accessed on 20 November 2013.

SOARES, N. B. **Educação ambiental no meio rural: estudo das práticas ambientais da Escola Dario Vitorino Chagas - comunidade rural do Umbu - Cacequi/RS.** Specialisation Monograph, Federal University of Santa Maria, Santa Maria, RS, 2007, 89p.

SOIFER, R. **Psicodinamismos da família com crianças:** terapia familiar com técnica de jogo. 2 ed. Petrópolis: Vozes, 1982.

TRISTÃO, M. **A educação ambiental na formação de professores:** rede de saberes. 2 ed. São Paulo: Annablume; Vitória: Facitec, 2008, 236p.

VEJABRASIL. Available at: <http://veja.abril.com.br/230909/lixo- domestico-reduzir-diminuir-impacto-ambiente-p-132.shtml> Accessed on 23 November 2013.

VIOLA, E. The ecological movement in Brazil (1974-1986): from environmentalism to ecopolitics. In: **Ecology and politics in Brazil.** PADUA, J. A. (org). Ecology & politics in Brazil. Rio de Janeiro: Espaço e Tempo: IUPERJ, 1987, 211p.

ZAKRZEVSKI, S. B. Environmental education in rural schools. In: BRASILIA, Ministry of Education, General Coordination of Environmental Education: Ministry of the Environment, Department of Environmental Education: UNESCO. **Let's take care of Brazil: concepts and practices in environmental education at school**, 2007. 248 p.

ZEPPONE, R. **Educação Ambiental:** Teorias e Práticas Escolares. 1ª ed. São Paulo: JM, 1999.

ANNEXES

Annex 1: Authorisation letter from the school

Secretaria da Educação e Cultura – 24ª DE
Escola Estadual de Ensino Médio Érico Verissimo
Portaria/SE Nº 00122 DE 15-04-2000
DO 26-04-2000 – Altera Designação
Rua Izaltino de Oliveira, 164
Restinga Sêca – RS

Oficio nº 054 Restinga Sêca, 17 de Outubro de 2013.

Prezado(a) Senhor(a):

Na oportunidade em que cumprimentamos vossa senhoria, vimos autorizar Simara Saquet Schio a utilizar os dados da escola para a pesquisa intitulada "Uma inferência acerca da conscientização ambiental de famílias rurais e urbanas ligadas à Escola Estadual de Ensino Médio Érico Veríssimo (Restinga Sêca, RS, Brasil)", cuja orientadora Jumaida Maria Rosito para o Curso de Especialização em Educação Ambiental.

Sendo o que tínhamos para o momento subscrevemo-nos atenciosamente.

Cleci Elia Borchardt
Diretora Id. 1453610/02
Cleci Elia Borchardt
Diretora
ID: 1453610/02

Annex 2: Questionnaire applied to students at the Érico Veríssimo State High School in Restinga Sêca, RS, Brazil.

FEDERAL UNIVERSITY OF SANTA MARIA CENTRE FOR RURAL SCIENCES
SPECIALISATION COURSE IN ENVIRONMENTAL EDUCATION
PROJECT: A STUDY ON THE LEVEL OF ENVIRONMENTAL AWARENESS OF RURAL AND URBAN FAMILIES CONNECTED TO THE ÉRICO VERÍSSIMO MIDDLE SCHOOL IN RESTINGA SÊCA, RS PROFESSOR: Simara Saquet Schio

ORIENTATOR: Profª. Dr Jumaida Maria Rosito

Answer the following questions as conscientiously as possible. The first four questions are aimed at characterising your family group and the collection of household waste where you live; the others are specific to the treatment of waste in your home. Please note that the form does not ask you to identify yourself.

1. Do you live in: rural area ( ) urban area ( )

2. How many people live in your house (INCLUDING YOU)? ( ) 2 to 3 ( ) 4 a 5 ( ) 6 a 7 ( ) 8 or more

3. Where you live, is there regular rubbish collection by the relevant municipal authorities?
Yes ( ) No ( )

4. Where do you live, is there any regular separate rubbish collection?
( ) Yes ( ) No ( ) There is, but it's not regular
Comments:

Now, read the following questions and mark with an "X" the alternative that best describes the practices related to household waste disposal in your family.

5. Do you sort your rubbish at home?
( ) Yes ( ) No ( ) Sometimes

If you said "Yes" or "Sometimes", answer the next question; if you said "No", ignore it and return to in question 7.

6. The separation is between organic waste (material of biological origin, such as food and drink waste, dead plants and animals, as well as wet paper) and inorganic waste (dry paper, plastics, glass, ferrous and non-ferrous metals).

( ) Dry rubbish and other types

( ) Other categories (please specify)

7. Do people in your household take care to use natural products and/or products that don't generate disposable waste, such as plastic or metallised packaging?

( ) Yes ( ) No

8. What do you do with your used cooking oil? How is it disposed of?

( ) In the kitchen sink

( ) On the ground

( ) Stored for later disposal at accredited sites.

( ) Other destination. Which destination? ______________________

9. To dispose of medicines (with an expiry date, or unused leftovers), your family is in the habit of:

( ) Add them to the regular rubbish

( ) Flush them down the toilet (in the case of liquids or tablets)

( ) Throw them in the sink/tank

( ) Deposit them in places accredited for this purpose

10. How your family gets rid of its rubbish:

( ) buys rubbish bags ( ) use carrier bags

( ) discard without packaging ( ) burn

11. Does anyone in your family smoke?

( ) Yes ( ) No

12. If the answer to the previous question is YES, what is the destination of cigarette butts? They

are usually

(   ) thrown in the yard or on the street (   ) placed in rubbish bins (   ) discarded without criteria, i.e. anywhere ( ) I don't know

13. In general, does your family believe that the correct collection and storage of rubbish can make a big difference to the health of the environment?

( ) No (   ) Yes

Comments:

Printed by Books on Demand GmbH, Norderstedt / Germany